Copyright © 2025 by Educate Learners

Published by Educate Learners

All rights reserved. No part of this publication may be reproduced, distributed, or transmitted in any form or by any means, including photocopying, recording, or other electronic or mechanical methods, without the prior written permission of the publisher, except in the case of brief quotations embodied in critical reviews and certain other noncommercial uses permitted by copyright law.

First Printing, 2025.

ISBN: 978-1-951573-49-2

www.educatelearners.com

A circle has

0 sides.

What is shaped like a circle?

A triangle has

3 sides.

What is shaped like a triangle?

This is a

square.

A square has

4 equal sides.

What is shaped like a square?

This is a

rectangle.

A rectangle has

4 sides that are 2 different lengths.

What is shaped like a rectangle?

This is a

oval.

An oval has

0 sides.

What is shaped like an oval?

This is a

diamond.

A diamond has

4 sides.

What is shaped like an diamond?

This is a

trapezoid.

A trapezoid has

4 sides.

What is shaped like an trapezoid?

This is a

pentagon.

A pentagon has

5 sides.

What is shaped like an pentagon?

This is a hexagon.

A hexagon has

6 sides.

What is shaped like an hexagon?

This is a

octagon.

An octagon has

8 sides.

What is shaped like an octagon?

This is a

heart.

A heart has

2 sides.

What is shaped like a heart?

This is a

crescent.

A crescent has

2 sides.

What is shaped like a crescent?

A star has

10 sides.

What is shaped like a star?

An arrow has

7 sides.

What is shaped like an arrow?

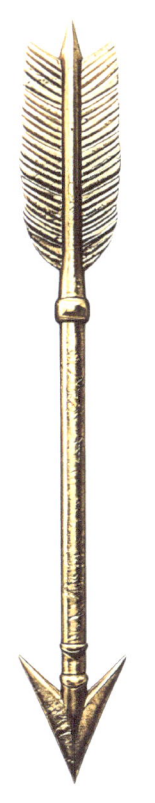
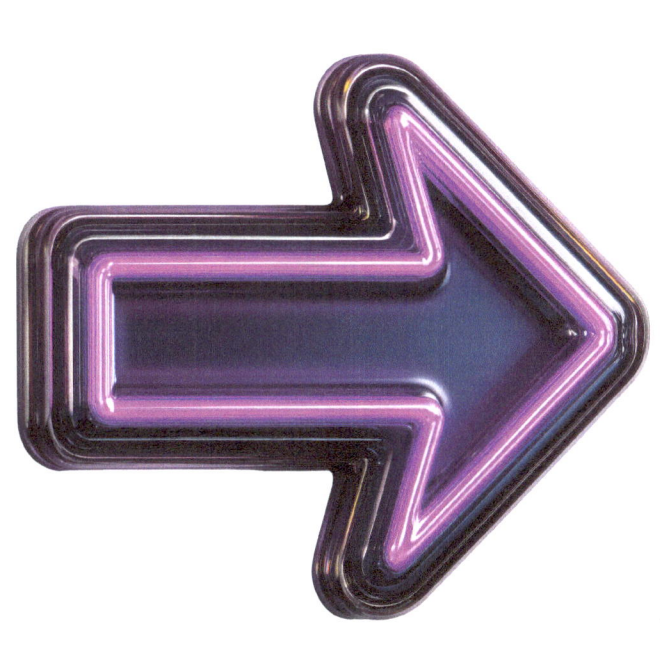

Thank you for reading!

Get a free year long subscription to our online education resource library when you purchase any one of our books.

Code: EDBOOKS

educatelearners.com

www.ingramcontent.com/pod-product-compliance
Lightning Source LLC
Chambersburg PA
CBHW041602070526
44586CB00003BA/51